50

essential facts
about
CLIMATE
CHANGE

by Gloria Barnett

Footprint to the Future

contents

Fact

1

Who is responsible for Global Warming?

Global warming is when the Earth's temperature goes up. Different places are having more heat waves and rain than before. This is happening more quickly than it used to because of human activity. We have been burning fossil fuels since the Industrial Revolution. This causes carbon dioxide to go into the air and trap heat, making the Earth hotter.

Carbon dioxide is causing most of the warming. It can take thousands of years to go away. Methane is causing the rest of the warming.

We must act quickly to keep the Earth's temperature from increasing too much. This is important so that all living things can survive.

Fact

②

Greenhouse Gases

Greenhouse gases are present in the Earth's atmosphere and contribute to the rise of our planet's surface temperature. When the Sun supplies light and heat to the Earth, that light and heat can either be reflected back into space or absorbed by the planet's surface. If there is an excess of carbon dioxide in the Earth's atmosphere, heat is trapped in the atmosphere, much like how a greenhouse reflects the light but absorbs the heat. The accumulation of carbon dioxide in the atmosphere creates a similar 'greenhouse effect' that warms the Earth. Due to human activities like burning fossil fuels over nearly 300 years, atmospheric methane concentrations have increased by over 150% and carbon dioxide by over 50%. These high levels of gases in the atmosphere have not been seen in over 3 million years.

Fact

③

What is the Carbon Cycle?

Carbon is essential for all life on Earth, forming complex molecules like proteins and DNA. It is found in the atmosphere as CO_2 and regulates the Earth's temperature. Carbon moves through a closed system known as the carbon cycle, circulating between the atmosphere, Earth, and living organisms. Most carbon is stored in rocks and sediments, while the rest is in the ocean, atmosphere, and living organisms. Human activities like burning fossil fuels have significantly impacted the carbon cycle, producing more CO_2 than at any other time in the past 3.6 million years.

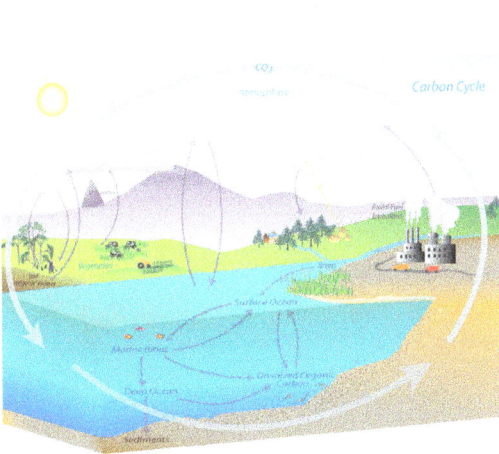

Fact

4

Carbon Dioxide

Carbon dioxide (CO_2) is an acidic chemical gas composed of one carbon and two oxygen atoms. It was studied by the Scottish scientist Joseph Black in the 1750s.

Plants absorb carbon dioxide from the atmosphere to use in photosynthesis. The reverse reaction happens in the cellular respiration process in animals when carbon dioxide is produced as a waste product. CO_2 travels in the blood from the body's cells to the lungs, where it is breathed out.

CO_2 is produced and released into the atmosphere when humans burn fossil fuels.

Fact

5

Oceans - Water - Rain

Oceans play a significant role in the water cycle of our planet. Water molecules comprise two hydrogen and one oxygen atom, known as H_2O.

The warmth from the Sun heats the surface water of the ocean, causing the seawater to evaporate and form clouds. These clouds carry individual hydrogen and oxygen atoms in the form of water droplets which travel around the world in the atmosphere. When the clouds cool down, the hydrogen and oxygen atoms combine to form water again, resulting in precipitation or rain.

Unfortunately, burning fossil fuels releases carbon dioxide, which, in turn, deposits excess heat into the atmosphere. The oceans absorb this extra heat, causing the water to become hotter and create more clouds and rain.

When it rains heavily, rivers overflow, causing damage to homes and businesses.

Fact

6

Ocean Warming Creates Global Heat

Scientists have found that the heat generated by global warming in the atmosphere is absorbed by the oceans, resulting in the highest-ever recorded ocean temperatures.

This extra heat causes more water to evaporate and turn to rain, affecting the animals living in the ocean. For instance, the water in the oceans has become so hot that it is causing the death of animals such as corals.

Fact

7

Oceans – Wind – Hurricanes

When oceans absorb more heat, the gases above the surface also get warmer. These hot gases rise, creating space at the ocean surface for strong winds to form.

The hotter oceans now influence weather patterns worldwide as strong winds form hurricanes. An increased number of storms is another aspect of climate change being observed globally.

Fact

8

Melting Ice at the North Pole

Ice naturally forms in the polar regions due to extreme cold temperatures, but the ice is melting rapidly. Global warming is causing climate change, even in these cold regions. The ice in the Arctic Ocean is melting rapidly. On a positive note, this melting in the Arctic will not result in increased sea levels worldwide since the ice at the North Pole is not on land but is already submerged in the sea.

When ice is present, its whiteness reflects the Sun's warmth back into space. However, as the ice melts, it reveals darker-coloured ocean water, which absorbs more heat and accelerates the melting process and global warming. The effects of global warming at the North Pole are already visible, and the ice is melting much faster than previously predicted. The Arctic's sea ice was recorded as the lowest ever in November 2023. Experts predict that all the ice will have melted in the Arctic Ocean by 2050.

Fact

9

Animals at the North Pole

The North Pole is the natural habitat of polar bears, the largest mammals on land. However, they can only survive by walking out onto the ice and hunting for seals. If the polar ice continues to melt, polar bears will be unable to hunt for food and will eventually starve to death.

Other animals, such as reindeer herds, are also experiencing food shortages due to the melting ice and snow in the regions surrounding the North Pole. This will lead to a bleak future for the polar north, losing biodiversity and animal life in vast quantities.

Fact
10

Methane at the North Pole

The land bordering the Arctic Ocean is covered by snow and ice, but a massive amount of tundra is underneath the ice. This tundra is made up of permanently frozen, black, sticky soil that contains millions of tons of methane gas. If the tundra becomes exposed to global warming, it will thaw and release methane gas. Methane is a potent greenhouse gas that can significantly threaten the balance of gases in the atmosphere.

The melting of the ice and snow in the North Pole is a danger to the survival of polar bears and other animals, causing severe implications for life on our planet.

Fact

⑬

We need to act now to stop the decline of freshwater ecosystems

Freshwater ecosystems (rivers, streams, lakes, wetlands, and underground storage known as aquifers) provide only 0.01% of the Earth's ecosystems. However, these ecosystems provide essential resources such as food, water, and energy to billions of people. Freshwater ecosystems contain one-third of all vertebrate species, with fish making up 40%.

Unfortunately, many freshwater ecosystems suffer from chemical, waste, and plastic pollution, overfishing and power generation. Scientists have reported an average decline of 35% within natural inland wetlands since 1970.

Freshwater is essential to support life on Earth, and with increasing drought being a symptom of climate change, urgent action is needed to preserve and sustainably manage freshwater ecosystems to prevent further harm. Helping freshwater conservancy can be an important area for volunteers. Why not find out where you can volunteer in your local area?

Fact

14

Protecting the Planet

Our number one task is to reduce the carbon dioxide in our atmosphere. The best way to do this is to stop using fossil fuels so global warming does not worsen.

We can all work together on this - by individually ensuring that we have the least possible CO_2 emitted from our homes and way of life. We can also help educate others about the problems caused by fossil fuels and share our successes in combating their use.

However, we must also look at ways to remove excess CO_2 from the atmosphere. Scientists are working on storing carbon dioxide, known as carbon capture, but the technology for this storage needs to be improved before this is an honest answer to the problem. Look out for any announcements in the future because if scientists can find an answer, we can start to clean up the atmosphere and stop global warming.

Fact

(15)

Importance of Seagrass

Oceans produce 80% of our oxygen from photosynthesis in seagrass and algae. To keep our oxygen levels secure, we must work harder to protect the seagrass and seaweeds in the oceans. We need laws to reduce the types of fishing vessels that damage both the seagrass and kelp seaweed that grows on the seabed. Marine parks are one way of doing this. Countries can increase the area of these protected areas and totally ban fishing within them. Water sports should be sectioned off away from areas containing seagrass. Once these marine parks become law, it usually only takes a few years before the kelp grows and seagrass meadows flourish again.

Fact
16

Volunteering to plant seagrass

All around the world, volunteers are helping to restore areas of the seabed where seagrass has been damaged. Research your local area - is there a marine park near you? Find your nearest seagrass planting project and see if you can help.

If not, write to your local council or your Member of Parliament to ask for help to protect the coastline. State the importance of seagrass for oxygen production. Get a group of people together to demonstrate peacefully to draw attention to the importance of keeping the ocean healthy.

Fact

17

Community Challenges

You don't have to do everything yourself – that would be almost impossible – but you can organise or join existing groups of people to help with nature projects that have already started worldwide. Community projects such as planting trees and seagrass are just two areas where you could help. Another area where local communities can act is rerouting rivers in river flood zones. With all the excess rain in recent years – rivers are flooding more often and causing problems in populated areas when rivers break their banks.

Community projects where lots of people volunteer to help are the way to get things done. Governments need to work more quickly; individuals can't do enough to make a difference, but communities working together is a way forward, with loads of people becoming volunteers to act on 'restorative' nature projects.

Research 'nature', 'wildlife trust', 'volunteering', 'national trust', or 'nature conservation' – to find somewhere you can help in your locality.

All the local work adds up to essential results around the world, so 'Think Global, Act Local' – you may be able to find work experience placements or weekend activities if you search your local area.

Fact

⑱

The job of Oceans is to keep the planet in balance

Oceans do an excellent job of producing oxygen from photosynthesis in marine algae and sea grass, then releasing the oxygen into the atmosphere. Oceans also absorb vast quantities of CO_2 from the atmosphere into the oceans. The oceans balance these two gases as part of the carbon cycle.

The ocean currents carry heat around the world. However, the global warming created by humans is warming the ocean and upsetting the natural balance of heat around the world. Hotter oceans are killing corals and damaging the food chains which are essential for life on Earth.

Humans need to help the oceans do their prime job. We need to stop polluting the oceans and help them keep the Earth's natural balance healthy.

Fact

(19)

Pollution in the Ocean

I'm sure we have all heard about pollution problems caused by plastics in the ocean and how discarded fishing nets can kill marine creatures by entanglement.

However, all sorts of other waste are thrown into the oceans, and humans must ensure that all waste is recycled or disposed of correctly.

Humans must stop dumping poisons into the ocean, too. Poisons such as mercury, industrial waste, radioactive materials, and sewage are all dumped into the oceans.

Dumping waste and poisons is not caused by climate change but by stupid humans!

Fact
20

Protecting animal species

Oceans are suffering from other problems humans cause - such as the reduced diversity of the oceans. Everything we do has an effect on the oceans. If we lose one species - it affects the whole food chain.

It has recently been discovered that looking after whales is essential to keeping the ocean clean - as whales help to create a balance in the lives of marine creatures. Whilst the whales eat smaller creatures, they also make large amounts of poo. It seems that their poo acts as a natural fertiliser which helps plankton grow, which feeds the animals that eat the plankton.

We need to educate everyone about the importance of oceans on this planet.

Fact

21

How can young people contribute to saving the planet?

It is important to acknowledge the significance of school nature projects. When students are motivated to undertake local initiatives, such as cleaning up beaches or planting trees, they tend to attract the attention of others. Adults can learn from the students when schools involve parents and the local community in these projects.

The enthusiastic attitude towards nature and caring for the planet encourages students to participate in community projects. It is a great event when children educate adults, and everyone enjoys themselves while feeling good about contributing to such an essential cause.

Fact

22

Who to listen to?

It's essential to be cautious about the information you consume, as there is a lot of it. Overloading yourself with negative news can harm your mental health, so it's important to stick to basic, reliable information. But who can you trust to provide you with that reliable information? One great source is Greta Thunberg, whose book "No One is Too Small to Make a Difference" is a must-read. You can also research and look up world leaders such as the United Nations General Secretary, Antonio Guterres, and US climate specialist John Kerry, who speak eloquently about climate change.

It's a good idea to find information about the good news, too - the changes already taking place and how sensible humans are at volunteering and helping fight climate change worldwide. I recommend you read the information on the WWF website.

https://www.worldwildlife.org/stories/the-good-news-about-climate-change

GRETA
THUNBERG

NO ONE
IS TOO SMALL
TO MAKE
A DIFFERENCE

Fact
23

Climate Research Organisations

There are many environmental organisations all over the world, which is excellent. However, choosing the one that suits you best can be challenging. Some organisations, like Greenpeace, request donations to fund their work, but many others don't require payment.

For young people, the most popular organisations are:

- **Young People's Trust for the Environment**
  (ypte.org.uk)

- **WWF**
  (www.wwf.org.uk/get-involved/schools/resources/climate-change-resources)

- **UNICEF**
  (www.unicef.org/environment-and-climate-change/youth-action).

Additionally, there are more options for people of all ages at: www.vox.com/future-perfect/2019/12/2/20976180/climate-change-best-charities-effective-philanthropy.

United Nations Information

FACT: Clean energy technologies produce far less carbon pollution than fossil fuels

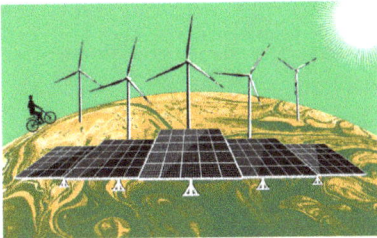

Fact

24

Check the science

If you enjoy reading and want to learn more about the impact of climate change on our planet, check out the online Global Climate journals from Wiley. They offer various journals on Biology and BioEnergy. You can visit (https://onlinelibrary.wiley.com/journal) to find a list of informative articles covering all aspects of life on Earth that are affected by climate change. Although these reports are not intended for children, try reading them to see if you can understand some scientific research being conducted and reported here. There is a lot of information available, so be careful not to overwhelm yourself. Choose an area that interests you and try to understand how much scientific research is already being done. It's exciting to see how hard scientists are working to help humans deal with the problems we have created.

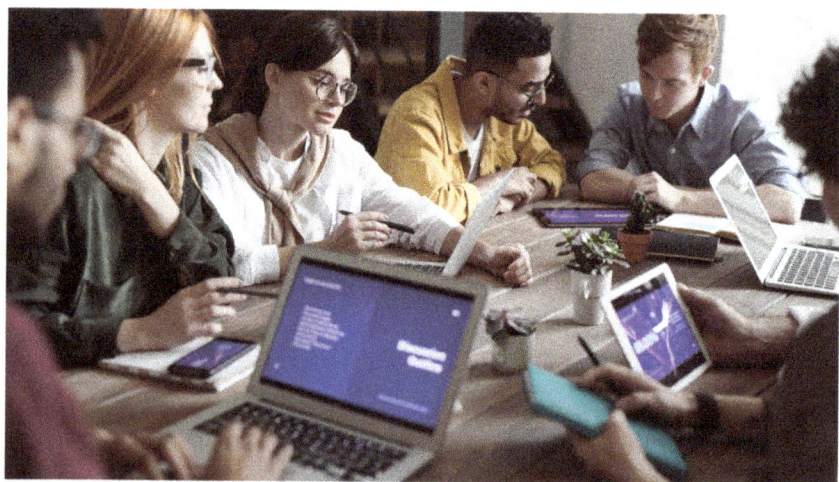

Fact
25

Follow your Passions

Much scientific evidence relating to climate change is highly technical and written by individuals in universities and research institutions. Therefore, focus on one specific area of climate change you are passionate about and write about it in your own words. For example, if you are an animal lover, consider researching and writing about how climate change affects animals and the efforts made to protect them in recent years. Sharing your knowledge and passion with others can help them better understand these issues, leading to more effective collective action against the devastating impacts of climate change.

An example of this is Greta Thunberg. She continues to advocate for people to stop using fossil fuels without complicating her mission with all the other different areas of climate change. Thunberg has identified a major problem that can have a massive effect on the effects of climate change. Therefore, she continues spreading her message to 'stop using fossil fuels'. It's likely she will only abandon this basic message once everyone on the planet follows her request.

Fact
26

Complain, protest and fight

Don't worry; you can assure your parents I am not suggesting you participate in violent protests. Your fight should be fought with words, not punches. When you stand up for what you believe in, it shows that you are passionate about it. Choose an area you want to specialise in and make your voice heard. There are many ways to do this without resorting to physical altercations or blocking roads.

The sensible way forward is to spread your message peacefully. Write to local businesses, councils, your local MP, politicians, local radio, and press. Ask them what they are doing to address your chosen issue, and keep following up until they take action. Look for answers from business owners that demonstrate they are committed to solving the problem, or ask politicians to debate the issue in Parliament and set a timeline for resolving it."

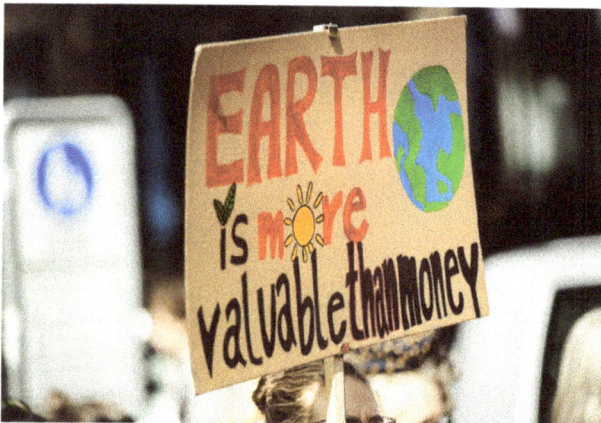

Fact

27

Your advice to the Government

Citizens in the UK elect Members of Parliament who represent different political parties. Once elected, Members of Parliament are expected to follow their party leaders. However, citizens need to encourage their representatives to support 'green' issues related to climate change. We must urge our politicians to act in the best interests of everyone instead of just following their parties or individuals in Parliament. Politicians must take their responsibilities seriously and look towards the planet's future.

Fact

28

Fight for what you believe in

We can all speak up for the causes we believe in, whether advocating for clean and healthy oceans, protesting against overfishing and super-trawlers, or fighting against the continued use of fossil fuels and plastics. While we may not be able to tackle every issue, we all have the power to make a difference in some way.

Why not try to stop using any water from a plastic bottle? Then, influence others to do the same thing.

Did you know it takes 1.85 gallons of water to manufacture the plastic for the average commercial bottle of water? That is a crazy fact!

Fact

29

What do we believe in?

We all have different opinions on which issues should be tackled first. However, time is of the essence, and we must work together to prioritise our efforts. Some may argue for finding ways to prevent wars because global peace is crucial and because the resources freed up could be redirected towards solving climate change issues.

Protesting for peace has proved effective for many years, and people's voices have been heard.

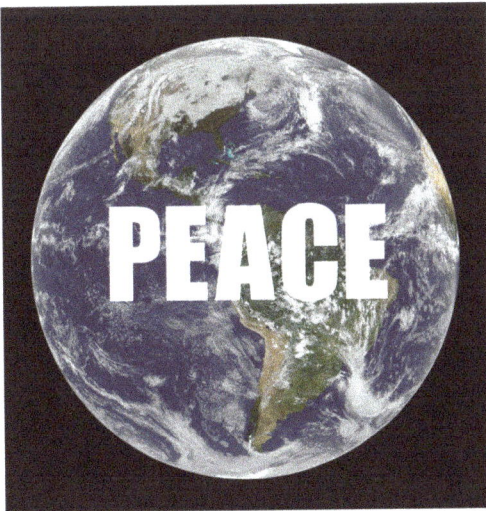

Fact

30

Understand and enjoy nature

Developing a love for nature can be one of the most valuable tools in a person's life. Whether you're taking a stroll in the woods or visiting the beach, immersing ourselves in nature can positively impact our mental health. If we fall in love with nature, we will be more inclined to learn about and protect the environment. This will benefit all our efforts to support the natural world and combat climate change.

If you live in a city, visit a local park, sit and listen to the birds, or watch the insects buzzing around a lake. Become immersed in your surroundings.

Fact
31

Recycle - Reuse

If you're new to recycling and reusing materials, you may feel overwhelmed about preventing plastic from polluting the oceans. However, it's simpler than you might think – all you need to do is get organised. Nowadays, most schools, colleges, and workplaces have separate collection bins for different materials. If there aren't any bins, you can speak to a teacher or boss and ask them to install them. Local councils then collect and sort these bins into separate recycling areas. Look around your home to see if you're doing everything possible to avoid using plastic.

Fact

32

Sustainability

The Earth has finite resources, and excessive consumption of these resources can deplete them faster than they can be replenished. This is an unsustainable practice that can cause long-term damage to the planet.

To achieve sustainability, we must use natural resources responsibly and carefully. This involves balancing meeting the present needs and preserving resources for future generations.

Environmental sustainability is an approach that emphasises protecting and preserving ecosystems and natural resources while minimising pollution.

Achieving sustainability requires collective efforts from everyone to create a world where people can work together in harmony with the Earth and everyone can thrive for generations to come.

Fact
33

Overshoot Day

Every year, humans consume more resources than the Earth can replenish? This day is known as Overshoot Day, and it signifies the point when humanity's demand for ecological resources surpasses the Earth's ability to regenerate those resources in a given year.

Overshoot day occurs earlier each year and is a significant cause for concern. It is unsustainable, and humans must change their ways to ensure a sustainable future.

What will happen when all of the Earth's resources are depleted?

Did you Know? Overshoot day for the year 2024 is projected to be on July 25th. From that day until the end of the year humans will be 'borrowing' resources from future generations. If there is to be a world which continues to supply resources to nourish all life on the planet, then we must become sustainable in our lifestyles.

Research? Go to **www.yearindays.com/earth-overshoot-day**

Fact
34

Consumerism

Consumerism is the idea that a person's happiness and well-being depend on acquiring material possessions and consumer goods.

While shopping might help improve mental health, buying things that are not necessary creates an excessive materialistic lifestyle, often leading to overconsumption and waste. Excessive waste harms the environment.

Therefore, ask yourself if you need to replace your clothing on a whim of the latest 'fashion' item. Become more creative with your wardrobe by reusing clothing and passing on children's outgrown clothing to another family. You can protect the environment and promote sustainable living if you are more creative with your wardrobe.
Remember, if you don't need it, don't buy it. Following this advice can help protect the environment and promote sustainable living.

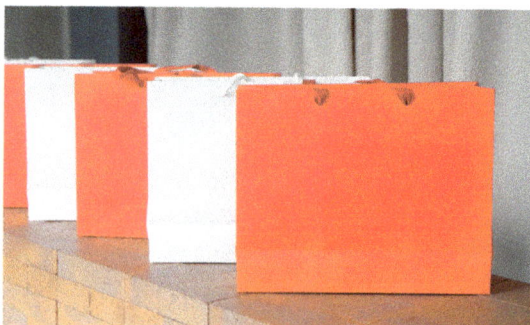

Fact

35

Reduce your transport carbon footprint

Your carbon footprint is the 'amount of CO_2 emissions released into the atmosphere due to your actions'. One way to reduce your carbon footprint while travelling is by walking or cycling, as these modes of transportation do not produce any emissions. Additionally, you can lessen car journeys or purchase an electric vehicle, which is emission-free. Taking a bus is also a good option as it allows you to share the cost of emissions with other passengers. Some cities now have electric buses available. If you must take a flight, check the airline's policy on reducing fossil fuels before booking. If you create carbon emissions, you can partially offset them by purchasing trees to be planted worldwide.

Fact

36

Reduce your food carbon footprint

If you think about the food you eat as part of a food chain, it's worth considering ways to shorten it. Let's take the following food chain as an example: Sun → Plant → Cow → Human. Every arrow in this chain represents the transfer of energy. The plant uses light energy from the Sun and CO_2 from the air to produce glucose through photosynthesis. When the cow eats the grass, it takes in the food energy from the plant. The cow uses some energy for its own bodily functions and produces meat, which is then consumed by humans. However, removing one element from this chain can reduce carbon output: Sun → Plant → Human. In this case, humans consume energy from plants rather than animal products. Therefore, it's worth considering reducing your meat consumption to help reduce carbon output.

Fact
37

Stop warming up the atmosphere

Do you enjoy spending time in the garden and having BBQs? Do you have a fire pit or a garden heater? It's essential to consider the impact of these activities on the environment. Burning charcoal on your BBQ or wood in the fire pit releases CO_2 into the atmosphere, contributing to climate change. It's worth checking where your electricity comes from, too, as it may be generated from burning fossil fuels. It's easy to overlook the environmental impact of our daily actions. Even small things like these can harm our planet.

Fact
38

Wasted Energy?

Have you ever noticed that some houses have snow on their roofs for a shorter period than others? This could be due to differences in insulation. When a home is poorly insulated, the warmth from inside will escape into the roof area, causing the underside to warm up and the snow on the roof to melt. Losing heat into the atmosphere leads to higher energy bills and an uncomfortable home. If you observe that the snow on your roof is melting faster than your neighbour's, it may be a sign that you need to improve your insulation. Doing so can keep your home warmer for a more extended period and save money on your energy bills. Improving your home's insulation is also essential to reducing your carbon footprint.

Fact
39

Installing Solar Panels

You may need to influence whoever makes the financial decisions in your home to consider installing solar panels. This is an obvious method for obtaining free energy. Sunlight is free!

We must take all necessary steps to reduce the use of fossil fuels. Although solar panels may seem expensive, the investment will pay for itself in the long run. Additionally, there are various deals available that you can take advantage of, such as selling excess electricity back to the power grid.

Using solar panels to power your home directly from the Sun is a great system, and you will be contributing to your local electricity supplier without burning any fossil fuels.

Fact
40

Join together as a community

Climate change requires collective action beyond individual efforts. Use your expertise to influence others and join forces with your community.

Collaborate with others to find solutions, such as building a solar panel array in a village, covering the cost with donations and generating income by selling excess electricity.

Even in a city, you can participate in community initiatives to promote a fossil fuel-free lifestyle.

Fact

41

Don't get into arguments

When people become aware of your knowledge about the possible future of planet Earth, you may encounter two opposing attitudes.

On the one hand, you may be pleasantly surprised by the enthusiasm and willingness of others to get involved. You can find people who are interested in sharing your knowledge and who will listen to you politely. You can initiate good discussions and ideas to move forward, and positivity will follow.

On the other hand, you may encounter rudeness and complete denial of the truth about climate change. In such a situation, it is best to consider how to deal with it in advance. Some people deny the existence of climate change because they are afraid of their future. You should not be surprised by how loud, rude, and noisy they can be.

To handle this situation, make a quick assessment: can you help these people become aware of climate change in a way that allows them to remain calm and envision a future? If that's the case, don't rush your knowledge; keep the discussion sensible. However, if they are adamant about their views, it is better to walk away and not engage in arguments.

Instead, show others how they can help by your actions. Your actions speak louder than words.

Fact

42

State evidence

Presenting accurate and truthful scientific facts when educating others about climate change is essential. Keeping notes on the scientific information you have learned and can share with others is also a good idea. Your objective should be to help others understand the future and make it your passion to pass on your knowledge. However, it is recommended not to overwhelm people with too many facts too quickly. Instead, be gentle and encouraging in your approach. Remember, helping others understand their future is the best thing you can do.

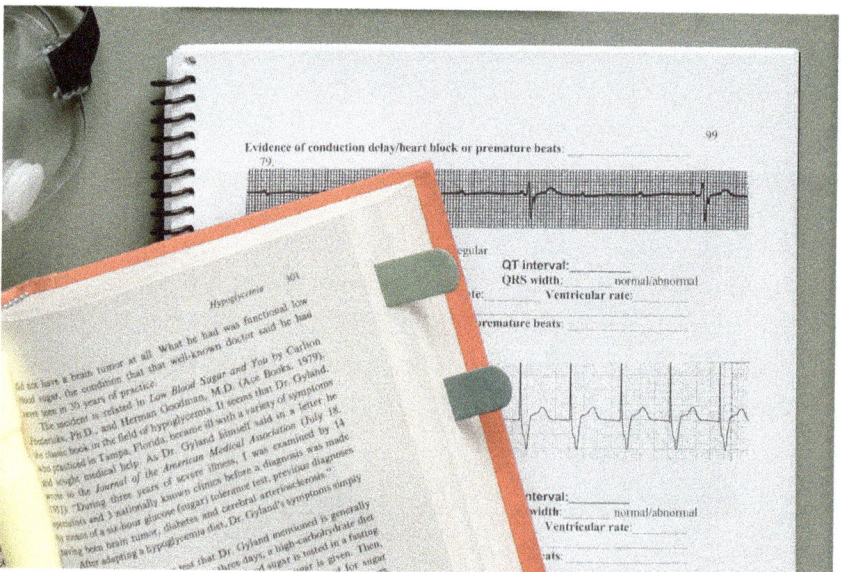

Fact
43

Inform others of local action

If you want to contribute to your community, listen out for any events. For example, there might be a beach clean-up or a peaceful protest outside a Member of Parliament's office with banners displaying your cause. Alternatively, there might be a local meeting to discuss a community project on sharing energy resources. If you are organising such an event, consider volunteering to help ensure everyone knows the date, time, location, and the invited speakers.

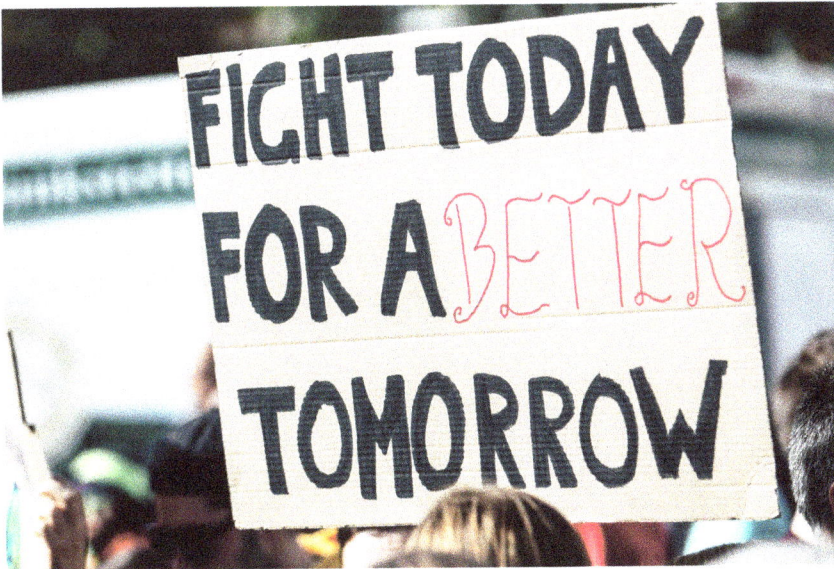

Fact
44

Inform people of good news stories

It's important to use the internet to research climate change and stay up to date, but try not to overwhelm yourself by reading or listening to everything. Take valuable notes and avoid relying solely on your memory.

Be aware of your mental health, and don't take on too much at one time.

When you come across good news related to climate change, please note it and share it with others. You can contribute to tackling climate change by raising awareness among your peers and beyond.

Listening to others' opinions, participating in discussions, challenging their approach, and demonstrating your understanding is vital.

To make things easier, encourage others to listen to your favourite blogs or podcasts and spread positivity in any way possible. Look for projects that fit the "If we do this now, then we can benefit XYZ in the future!" approach and get others involved.

Fact
45

Become a voice to be listened to

Consider making yourself a person worth listening to, someone who can provide valuable information that people want to hear. You could create a blog or podcast to share your researched and well-understood positive news with the public.

Moreover, sharing information about 'The Earthshot Prize, an initiative to recognise and reward environmental solutions, can help people understand worldwide encouraging news. By doing so, everyone can be hopeful for the future of our planet. Here's the website URL for more information: https://earthshotprize.org

Fact
46

Do your best in education

If you hope to earn respect for your knowledge and opinions, it's crucial to prioritise your education. Rather than solely focusing on saving the planet, it's important to excel in your exams and progress through different stages of education. This will equip you with the necessary tools to tackle future challenges.

Consider what kind of work would suit you as a future leader in climate change. You could be a conservationist, work alongside scientists to find alternatives to fossil fuels, or even become a leader in climate change discussions and writing. Science is making significant advancements in all areas required to combat the rise in temperature in the atmosphere and oceans. Everyone has unique talents, so choose an area of work that resonates with you.

Fact
47

Be an Influencer

Becoming an influencer is a relatively new trend, which refers to young people who use social media and podcasting to showcase their opinions to the world. However, age can be a factor in sharing knowledge about climate change and the planet's future. Older people may have different levels of respect for the opinions or methods of younger influencers. Other generations also rely on various sources of information, such as newspapers, magazines, TV, or radio stations.

One way to better understand these differences is to conduct surveys across various age groups. By asking questions about where people get their information and what they know about climate change, we can comprehend what information is required for different age groups.

As an influencer, you need to inform and promote action against future climate change problems.

Fact
48

Being a Future Leader

Becoming a leader in the fight against climate change requires a comprehensive understanding of several key areas. These areas include the following:

1. The complex scientific principles behind climate change.
2. The ability of humans to change their lifestyles and adopt sustainable solutions.
3. The timeline for implementing proposed solutions.
4. The intricate economics of a complex world, where phasing out one type of problem may cause financial dilemmas such as job losses or food shortages in other areas or countries.
5. How to assist poorer countries and those facing sea level rise.
6. Finding the finances to pay for the necessary changes.
7. Helping humans adjust to changes and be prepared to tackle the challenge of doing things differently.
8. Allocating government funds to different areas of importance and prioritising climate change at the top of the agenda for the future.
9. Acknowledging that climate change will not destroy the planet, but it endangers the existence of all life on Earth.

Fact
49

Being passionate

However you choose to help life on Earth continue to flourish, and however much you may worry about the future of human beings, know that there is hope for a better tomorrow if we follow these steps:

1. Educate yourself and others.
2. Understand the symptoms, causes, and cures for climate change.
3. Learn how individuals can help the planet.
4. Collaborate with groups of committed citizens.
5. Share your knowledge with commitment and passion.
6. Take a peaceful approach to fight for what you believe in.

By taking these statements into consideration, the future looks brighter. The Sun has a few billion years of life left and will continue to shine on us all.

The author of this book has been a passionate advocate for the protection of the natural environment for many years.

Fact
50

Become a Planetary Citizen

Working alone will never solve our climate problems, and relying on organisations and governments is too slow. Working as a group in a community is the best we can do.

What is the mission of Planetary Citizens?

To promote global unity and sustainability through diverse, holistic approaches to educational and environmental challenges, fostering regenerative cultures and democratic communities for a harmonious future.

In 1978 - Margaret Mead, an anthropologist, was awarded the prize of 'Planetary Citizen of the Year' – this is her famous statement:

"Never doubt that a small group of thoughtful, committed citizens can change the world."

You are a member of that small group ... go out and change the world!

Glossary

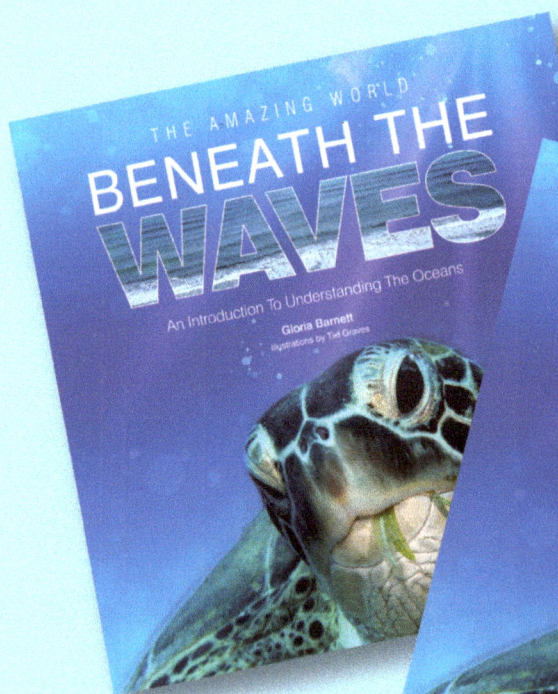

More books to enjoy

Fiction

Fiction

Children's books

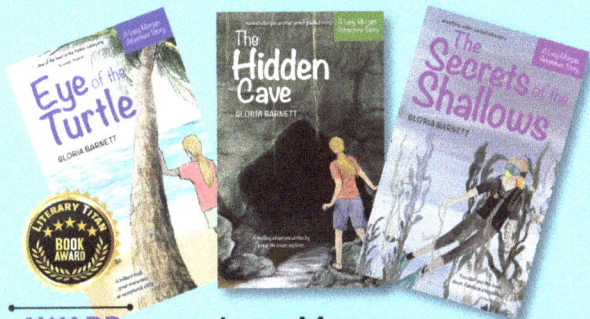

Eye of the Turtle
GLORIA BARNETT

The Hidden Cave
GLORIA BARNETT

The Secrets of the Shallows
GLORIA BARNETT

LITERARY TITAN
BOOK AWARD

AWARD WINNING AUTHOR

Lucy Morgan adventure series

Age 8-12+

Non-fiction

Guides to understanding the Oceans

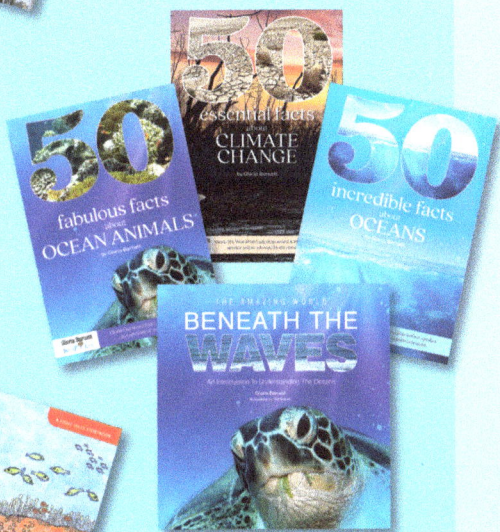

50 fabulous facts about OCEAN ANIMALS

50 essential facts about CLIMATE CHANGE

50 incredible facts about OCEANS

THE AMAZING WORLD
BENEATH THE WAVES
An Introduction To Understanding The Oceans

Fiction

Children's books

Prickle the Puffer Fish

Ravi the Ray

Logan the Lobster

Fishy Tales storybooks

Age 3-6

Gloria Barnett is an award-winning author with a passion for science, particularly in ocean studies and climate change. Her extensive knowledge and expertise in these fields make her an exceptional writer for all age groups.

Gloria's love for the natural world is infectious. Her 30 years of experience exploring the world's oceans as a master scuba diver, sailor, and underwater videographer has given her a practical understanding of the sea and its inhabitants.

She has been a strong advocate for conservation for many years and written a range of books, including non-fiction guides on oceans and climate change, as well as children's fiction for 3-6 and 8-13-year-olds.

www.barnettauthor.co.uk.

Remember
- there is only one planet Earth
- it is our home - please help to look after it!

IMAGE CREDITS/FACT NUMBERS

NOAA	3, 18	United Nations	23, 50
BioplasticNews.com	4	National Geographic	24
G.Barnett	6, 13, 21, 40, 49	NASA	29
Adobestock	15		
Dr. Colarusso EPA	16		
Project Seagrass, Wales	17		
G.Thunberg	22		

All other images are from contributors to pexels.com

Thank you to everybody.

Published in 2024

An imprint from
Footprint to the Future
Publishers of resources to help understanding
of Planet Earth.
165 London Road Temple Ewell, Kent, CT16 3DA

A CIP record for this title is available from the British Library

ISBN: 978-1-7393084-1-4

Book Design: www.amberdesigns.com

Published by: www.footprinttothefuture.co.uk

www.ingramcontent.com/pod-product-compliance
Lightning Source LLC
Chambersburg PA
CBHW051504270326
41933CB00021BA/3470